Sandip Pal

Lidar and Radiometer Sounding of Atmospheric Aerosols and Trace Gases

AF153161

Sandip Pal

Lidar and Radiometer Sounding of Atmospheric Aerosols and Trace Gases

Using Wavelet-transform analysis for long-term climatology of aerosol content over an urban site (Pune) in India

LAP LAMBERT Academic Publishing

Impressum / Imprint

Bibliografische Information der Deutschen Nationalbibliothek: Die Deutsche Nationalbibliothek verzeichnet diese Publikation in der Deutschen Nationalbibliografie; detaillierte bibliografische Daten sind im Internet über http://dnb.d-nb.de abrufbar.
Alle in diesem Buch genannten Marken und Produktnamen unterliegen warenzeichen-, marken- oder patentrechtlichem Schutz bzw. sind Warenzeichen oder eingetragene Warenzeichen der jeweiligen Inhaber. Die Wiedergabe von Marken, Produktnamen, Gebrauchsnamen, Handelsnamen, Warenbezeichnungen u.s.w. in diesem Werk berechtigt auch ohne besondere Kennzeichnung nicht zu der Annahme, dass solche Namen im Sinne der Warenzeichen- und Markenschutzgesetzgebung als frei zu betrachten wären und daher von jedermann benutzt werden dürften.

Bibliographic information published by the Deutsche Nationalbibliothek: The Deutsche Nationalbibliothek lists this publication in the Deutsche Nationalbibliografie; detailed bibliographic data are available in the Internet at http://dnb.d-nb.de.
Any brand names and product names mentioned in this book are subject to trademark, brand or patent protection and are trademarks or registered trademarks of their respective holders. The use of brand names, product names, common names, trade names, product descriptions etc. even without a particular marking in this work is in no way to be construed to mean that such names may be regarded as unrestricted in respect of trademark and brand protection legislation and could thus be used by anyone.

Coverbild / Cover image: www.ingimage.com

Verlag / Publisher:
LAP LAMBERT Academic Publishing
ist ein Imprint der / is a trademark of
OmniScriptum GmbH & Co. KG
Bahnhofstraße 28, 66111 Saarbrücken, Deutschland / Germany
Email: info@omniscriptum.com

Herstellung: siehe letzte Seite /
Printed at: see last page
ISBN: 978-3-659-52998-6

Zugl. / Approved by: Department of Physics, University of Pune, Diss., 2004

Acknowledgements

With great pleasure, I express my deep sense of gratitude and heartfelt thanks to my guide Dr. P.C.S. Devara, Head, Physical Meteorology and Aerology Division, Indian Institute of Tropical Meteorology, Pune, for his untiring efforts, valuable guidance, keen interest and encouragement throughout the course of the project work. I have got a great experience throughout the period of my working with him.

I take this opportunity to thank Dr. P. Pradeep Kumar, Co-ordinator, M.Tech. (Atmospheric Physics), Department of Physics, University of Pune, Pune for permitting me to carry out the Project work at I.I.T.M., Pune. I also acknowledge the invaluable help that he had extended from time to time during the period of the work.

I put forth, with great reverence, my heart felt gratitude to Dr. P.N. Sen, Guest Faculty, Department of Atmospheric and Space Science (DASS), University of Pune, Pune for his advice and inspiration at various stages of the work.

My sincere thanks are due to Dr. G.B. Pant, Director, I.I.T.M., Pune for permitting me to carry out the work by providing the Library, Hostel and other facilities.

My thanks are also due to Mr. M.A. Kulkarni, J.R.F, Department of Atmospheric and Space Science, University of Pune, Pune for providing me invaluable help in this work.

I am thankful to S/Shri S. M. Sonbawne, S.K. Saha, K.K. Dani, A.K. Srivastava, Dr. R.S. Maheshkumar, M.I.R. Tinmaker and Miss. R. Bhawar for the pains that they have taken in helping me and especially grateful to them for their valuable advice and inspiration throughout the course of the work.

I would like to thank the Staff of the IITM and Jayakar Libraries, for providing me all their support and necessary books, journals and proceedings and a congenial atmosphere to study there.

I wish to thank the Officers of the Regional Transport Office (RTO), Pune and the Director of Maharashtra Chamber of Commerce for providing the Vehicle and Industrial growth information, respectively. Special thanks are due to the Officer-in-Charge of Zilla Parishad, for providing the population growth of the Pune city.

I express my sincere thanks to all the employees of IITM for presenting me a pleasant atmosphere during my stay at the Institute.

My elder brother Joydip, a fantastic guy, words can't express my gratitude to and love for him for his continuing encouragement. I am greatly indebted to my parents for their help, dedication and patience during this period.

Last but not least, I thank my entire friend circle for their encouragement, help and moral support.

SANDIP PAL

Table of Contents

Abstract

Over 700 weekly-spaced vertical profiles of aerosol number density have been archived during 14-year period (October 1986 – September 2000) using a bi-static Argon ion lidar system at the Indian Institute of Tropical Meteorology, Pune (18°43′N, 73°51′E, 559 m above mean sea level), India. The monthly resolved time series of aerosol distributions within the atmospheric boundary layer as well as at different altitudes aloft have been subjected to the wavelet-based spectral analysis to investigate different characteristic periodicities present in the long-term dataset. The solar radiometric aerosol optical depth (AOD) measurements over the same place during 1998-2003 have also been analyzed with the wavelet technique. Wavelet spectra of both the time series exhibited significant quasi-annual (around 12-14 months) and quasi-biennial (around 22-25 months) oscillations at statistically significant level. A brief overview on the lidar and radiometric data sets including the wavelet-based spectral analysis procedure is also presented. A brief statistical analysis concerning both annual and inter-annual variability of lidar and radiometer derived aerosol distributions has been performed to delineate the effect of different dominant seasons and associated meteorological conditions prevailing over the experimental site in Western India. Additionally, the impact of urbanization on the long-term trends in the lidar measurements of aerosol loadings over the experimental site is brought out. This was achieved by using the lidar observations and a preliminary data set built for inferring the urban effects of the city of Pune, which included population, number of industries and vehicles etc. in the city.

5

1. Introduction

Atmospheric aerosols are distinctly recognized as an important governing factor in many environmental aspects of climate and radiative forcing processes (Twomey, 1991), as well as in the health effects of air quality (Samet et al., 2000). For instance, several researchers investigated in detail the complex relationship between the increase/decrease of aerosols and associated decrease/increase ("Dimming"/ "Brightening") in surface solar radiation. Aerosols exhibit large spatial and temporal variability due to a variety of production, removal and transport processes. It is also due to their potential long-term changes in atmospheric radiation balance caused by natural variability in aerosol concentration, human-induced changes or sudden natural events such as volcanic eruptions (Andreae, 1995; Charlson et al., 1999). Thus, it is essential to monitor systematically the aerosol features particularly the vertical profiles of aerosol concentration and column-integrated aerosol optical depth (AOD) over longer time period. For instance, AOD serves as an excellent indicator of aerosol load in the atmospheric column. The knowledge of long-term variability in aerosols burden over a particular region is of paramount interest for studying their impact on climate. Even though numerous experimental and theoretical techniques have been developed during last about five decades, our ability to assess the impact of aerosols on Earth's environment is incomplete due to many inherent unknown complex properties of aerosols such as physicochemical and source-sink cycle transformation patterns including their direct and indirect effects (IPCC, 2001).

More than 20 % of the atmospheric aerosol mass is of anthropogenic origin. The role of the atmospheric aerosols within the climate system is hardly understood but certainly important. The increasing awareness that man is perturbing the atmosphere and encroaching on the environment on the global scale coupled with the complexity of the atmospheric phenomena and the limitation of our ability to understand them requires decisive action. As atmospheric aerosols show high degree of variability in space and time in their characteristics and as the anthropogenic share of total aerosol loading is quite substantial, it is essential to monitor systematically the aerosol features over longer time scales (Devara et al., 2001a). One such parameter that plays an important role in the study of aerosol forcing on climate system is the Aerosol Optical Depth (AOD).

Aerosols have a strong influence both on the terrestrial climate and on society. Aerosols modulate the Earth's radiation balance directly by scattering and absorbing incoming radiation (Charlson et al., 1992). By serving as cloud condensation nuclei, aerosols influence the number and size of cloud droplets. This process can affect cloud radiative properties (Twomey, 1974) and the lifetime and precipitation properties of clouds (Albrecht, 1989). Aerosols act as surfaces where important reactive photochemistry takes place, both in the troposphere (Dentener et al., 1997; Lelieveld et al., 1997), and in the stratosphere. They are believed to fertilize ocean biota by the transfer of soluble nutrients from land to ocean (Martin, 1990). Aerosols can also play a role in the health of biological organisms.

The most important aerosol species are thought to be composed of sulfate, carbon, mineral dust, sea salt, and nitrate. Aerosols are often observed to exist as mixtures of these species. While sources of mineral dust and sea salt are primarily natural in origin, the other species have both natural and anthropogenic sources. As human kind significantly changes the land surface properties, even mineral dust can be viewed as being anthropogenically influenced. In spite of intense study of the atmosphere's aerosol over the last twenty years, there are still large gaps and uncertainties in our knowledge about this important component of the Earth's climate system (IPCC, 1996).

Optical properties of aerosol particles are mainly dependent on the distribution of atmospheric temperature, therefore on stability, and the humidity due to their hygroscopic properties, while the kinematics of aerosols is solely governed by wind direction and speed and underlying orography or surface. The aerosol burden over a region is mainly triggered by aerosol emission and production while sedimentation and precipitation scavenging governs their removal mechanism. Synoptically driven aerosol transport and associated atmospheric circulation also play a key role in the aerosol removal processes.

Continuous monitoring of the vertical distribution of optical properties of aerosols like particle extinction coefficient and its column integrated values are considered to be important parameters to investigate aerosol removal processes influenced by both synoptic and large-scale processes (e.g., south-west monsoon system). On the other hand, quantification of optical properties is also necessary for modeling the impact of aerosols on climate. Additionally, this information can help to improve the existing models designed to study the urban aerosol dynamics (Ackermann et al., 1998).

LIdar (LIght Detection And Ranging) and RADiometric (LIRAD) techniques have been recognized to be powerful means for the study of aerosols over large atmospheric volumes with good spatial and temporal resolutions from a single surface location or over larger regions by means of airborne or satellite-borne systems. Both direct and remote sensing methods are used to determine the physiochemical properties of atmospheric aerosols. However, the latter has the advantage in providing range-resolved information on the physical properties of aerosols. Much of the contribution to this field has come from optical remote sensing methods (Killinger and Mooradian, 1983; Devara, 1998). Both lidar and radiometric technique are considered to be powerful means for studying aerosols over large atmospheric volumes with good spatial and temporal resolutions by using ground-based platforms (both stationary and mobile) or over larger regions by means of airborne or satellite-borne platforms.

Atmospheric aerosols reveal various interesting periodic cycles including classical diurnal, weekly, and seasonal cycles due to natural or anthropogenic influences or heterogeneity in the land-use pattern. The present study has been highly benefited from the long-term measurements of aerosols archived at Indian Institute of Tropical Meteorology (IITM), Pune. A complete set of systematic observations available during the period between 1986 and 2000 has been used keeping a focus in determining different scales or frequencies embedded in the time series is used. Key challenges lie then in exploiting the time series to detail the physical characteristics of aerosols and their variability in time. This calls for an advanced statistical analysis technique like wavelet-based spectral analysis (Foufoula-Georgiou and Kumar, 1995).

Studies of the oscillatory phenomena have traditionally been performed using fast Fourier transform (FFT)-based analysis (Cooley and Tukey, 1965). Fourier transform alone cannot provide a comprehensive description of the properties of the non-stationary processes because it yields a mapping that is localized in frequency but global in time. The wavelet analysis in contrast to the FFT allows one to localize irregularities both in time and scale (frequency) domains and then to resolve isolated features in a time series. A detailed comparison of Fourier spectra and wavelet spectra can be found in Hudgins et al. (1993).

The concept of wavelet transform (WT) was introduced in the early 1980s and then was used by many researchers for atmospheric and oceanographic studies. For instance, Farge (1992) applied WT analysis to study atmospheric turbulence while Meyers et al. (1993) applied WT to investigate the dispersion properties of Yanai waves. Torrence and Compo (1998) provided a detailed description on the technique and discussed its application in deriving the oscillations involved in El-Nino. Wavelet analysis has been also applied to study gravity waves observed with lidar and to detail the structure and dynamics of multi-layer Cirrus cloud field probed with radar (Quante et al., 2002).

Distribution of aerosols within and above the atmospheric boundary layer (ABL) has a significant impact on local pollutant budgets at a variety of temporal and spatial scales. Detailed insights into these scales can improve our understanding of the structure of the lower troposphere and aerosol processes taking place in there. The long-term data set is used to illustrate the type of the results that is expected from the localized space-scale analysis. The wavelet-based techniques fortunately are able to extract and analyze multi-scale features hidden in a time series.

In the present work, syntheses of multi-year aerosol data that have been archived using the lidar for 14 years and radiometer for 5 years at Pune, India have been presented. Retrieval of aerosol column content (ACC) was performed from the lidar vertical profile measurements of aerosol concentration. This study provided the longest data series of such ground-based measurements in South Asia. Particularly, statistical characteristics of systematic aerosol lidar observations performed during the long term period have been documented. This work mainly devotes to the application of continuous wavelet transform (CWT) to the aerosol measurements with an aim to resolve the localized signals and scale interactions. This helped determine the oscillatory behavior of the lidar-derived aerosol distributions and radiometer-measured AODs. By applying the CWT analysis to the time series of AODs at different wavelengths, determination of the nature of the dependencies of the oscillatory frequencies on different wavelengths, hence on the types of the aerosol particles is aimed. Additionally, following a subjective approach, an attempt has been made to investigate the impact of urbanization parameters such as industrialization, population and number of vehicles on aerosol burden over the site

In view of the growing importance of the knowledge of the atmospheric constituents and need of such information, particularly related to the long-term trends of the aerosol loading in a tropical semi-arid urban station, an attempt is made in this study to get a comprehensive picture of the lidar derived ACC (Aerosol Column Content) time series for a period of about 14 years and Microtops retrieved AOD, TCO and PWC data series for more than 5 years with the help of various statistical algorithms and Wavelet Transform Analysis.

2.   Experimental Site

2.1   Geography and Location

The prevailing environment over the experimental station Pune ($18°32'$N, $73°51'$E) is urban and the aerosol type present over the observational site is assumed to be a mixture of water-soluble, dust and soot-like aerosols. Soil-dust is the major source of aerosol pollutants present over the experimental station. This assumption is in moderate agreement with the results of lidar scattering multi-angle and multi-spectral data, which revealed the aerosol size index to be between 3.5 and 5.5 and refractive index to be between 1.4 and 1.6 (Devara et. al., 1995; Pandithurai et. al., 1996).

The experimental site in the Pune city is located at an elevation of 573 m above sea level (ASL), approximately 100 km inland from the west coast of India. Pune is located on the leeward side of Western Ghats (Sahyadri mountain range) which form a barrier from Arabian Sea (Figure 1). The city is surrounded by hillocks forming a valley-like configuration, with its tallest hill, Vetal Hill, rising to 800 m ASL; just outside the city, the Sinhagad fort is located at an altitude of 1300 m ASL. Pune is eighth largest metropolis in India and the largest city in the Western Ghats.

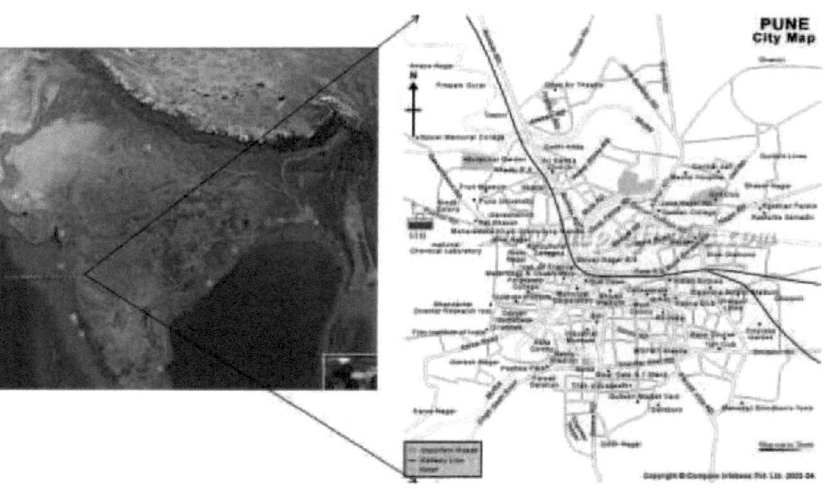

Figure 1: Location of the experimental site in Pune.

The aerosol type around the site is a mixture of water-soluble, dust and soot-like aerosols, of which soil-dust is the major source of aerosol pollution (Khemani, 1989). Formation of aerosols in the accumulation-mode is considered to be due to gas-to-particle conversion processes, whereas the coarse-mode aerosols are attributed mainly to wind-blown dust. Dispersion of pollutants, particularly those in the lower levels of the atmosphere above the site are believed to be affected by the circulation processes that evolve due to complex terrain of the experimental location.

Pune has a tropical wet and dry climate with average temperatures ranging between 20 and 28 °C. The summer monsoon lasts from June to September, with moderate rainfall and temperature ranging from 10 to 28 °C. An active monsoon spell is considered to be the potential aerosol removal mechanism for this region. Most of the annual rainfall in the city occurs during these months while July is the wettest month of the year. The weather at the experimental site during the pre-monsoon (March-April-May) is very hot and gusty winds introduce significant dust content into the lower troposphere. Winds are predominantly westerly during the summer monsoon that brings large influx of moist air from the Arabian Sea. This results in intermittent or continuous rain that makes the atmosphere relatively free from dust in this season. Normally, low aerosol loading is observed in this season (Devara et al., 2003). The westerly flow weakens in the lower troposphere and the easterly flow sets in during the post-monsoon (October-November) season causes rich nuclei of continental aerosols passing over the station. The clear-sky, very low humidity and light surface winds during the winter (December-February) season result in low-level inversions-caused dust haze during the morning and evening hours over the station. Intrusion of dry polar continental air mass in the wake of low pressure systems associated with the western disturbances moving across the north-western parts of India also occurs during this season (Khemani et al., 1995).

2.2. Urbanization aspects of Pune

The road and the railway connecting the two principal cities of Maharashtra have attracted industrial location almost as a ribbon development since independence (i.e. 1947). The approaches to Pune from Chinchwad onwards, mark the rise of a new industrial and complex that is a giant in the making with large factories producing machine and tools, electrical and electrical goods, plastics, pharmaceuticals, engineering parts. These sprawls out on what was once dry scrubland and find their location strength in the adjoining rail and roadways. Old Pune itself is less industrial and more residential, and one sees the unusual spectacle of commuters radiating out to all sides to serve in the new industrial units. This is in contrast with the normal patterns in which commuters from the urban fringe gravitate to a closely-knit industrial center. Modern industry finds its location in the Pune–Ahmednagar road (Electronics and Processed foods), the Pune-Solapur road (Engineering and Structural and Paper), Pune-Satara road (Engineering) and of course in the region of Pimpri-Chinchwad area. Also small-scale industries have a sporadic location. The major industrial base, however, lays all rounds on the fringe of the old city along the major routes with a heavy bias towards the Engineering industries. The total number of the industries of course shows a rising trends through-out the time scale from 1985-2002.

The growth in the motor vehicle population in Pune city determines the contribution of auto emissions to the overall air pollution in the city. As the pollution load attributable to the auto exhaust depends on the vehicles traveled and the growth with time therein, for assessing the need for improvement in the vehicular emissions, it is necessary to study the growth trends of motor vehicles. The vehicular emission in the Pune city are contributed not only by the private owned vehicle and private transport vehicles used by residents of the city, but also by the private and public transport vehicles commuting and passing through the city. The total number of vehicles in the Pune city during 1981-2002 showed an increasing trend. During this time interval, the population also showed an increasing trend in Pune city.

3. Instruments and measurements used

3.1 Lidar

The lidar system used consists of a continuous wave Argon-ion laser as transmitter and a 25-cm diameter Newtonian telescope with a focal ratio (focal length divided by the diameter of the telescope) of f/7.6 as receiver equipped with an intermediate optics assembly composed of condensing-collimating lenses, narrow-band interference filter and Peltier-cooled photo multiplier tube (Figure 2). The system is operated in bi-static configuration with its transmitter horizontally separated from the receiver by about 60 meters. The vertically transmitted laser beam (at wavelength of 514.5 nm with an average power of about 500 milli-watts) is scanned by the receiver at pre-programmed elevation angles corresponding to different altitudes. The complete system was installed on the terrace of the Institute building in order to get rid of the contamination of the signals from topographic targets such as tall buildings, trees etc.

3.2 Lidar data analysis

The lidar scattered intensity profiles obtained at the wavelength of 514.5 nm have been utilized to derive the profiles of aerosol number density following the inversion technique discussed in detail by Devara and Raj (1987), Devara et al. (1995a,b), and Raj et al. (1997). This technique has been extensively used for the analysis of lidar data obtained at IITM since 1986. This approach has been used in this study; its description is presented in discrete notation following Devara et al. (1995a). The profile of aerosol number density can be

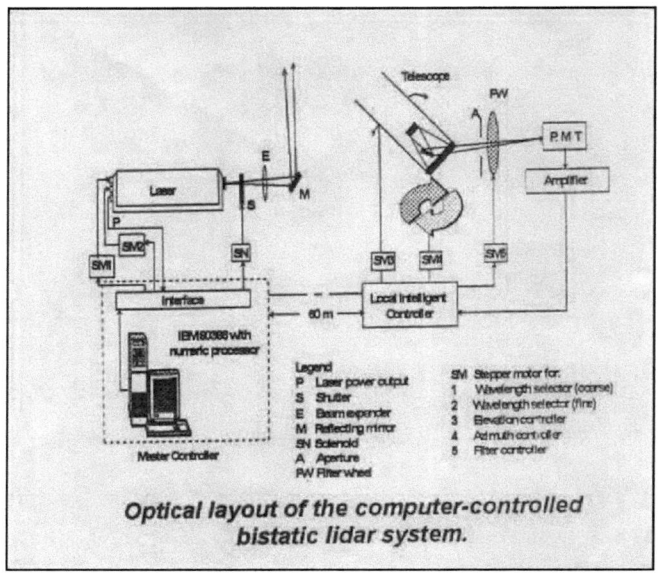

Figure 2: Optical layout of the CW Bistatic lidar system at IITM, Pune

computed from the background-corrected signal intensity, obtained at different altitudes by sampling the vertically transmitted laser beam at different elevation angles of the receiver, by using the following bi-static lidar equation:

$$P_R = \frac{P_T V T A_R \eta N \sigma(\theta)}{R_1^2 R_2^2 \, d\omega_1} \tag{1}$$

where P_T and P_R are the transmitted and received power, respectively; V is the scattering volume; T is the atmospheric transmittance along the transmitter–scattering volume–receiver path; A_R is the collecting area of the receiver; η is the system constant including the overall efficiencies of the transmitter and receiver optics; $\sigma(\theta)$ is the differential cross-section at scattering angle θ; R_1 and R_2 are the ranges from the transmitter and receiver respectively to

16

the center of scattering volume; $d\omega_1$ is the solid angle of the transmitting beam. The total scattering coefficient term $N\sigma(\theta)$, which is sum of the molecular and the particle backscatter coefficients can be expressed as

$$N\sigma(\theta) = N_a\sigma_a(\theta) + N_m\sigma_m(\theta), \tag{2}$$

where N_a and N_m are the concentrations and $\sigma_a(\theta)$ and $\sigma_m(\theta)$ are the differential cross-sections at scattering angle θ for aerosol particles and molecules, respectively. The molecular number density N_m is determined from pressure and temperature profiles obtained from local radiosonde measurements or following classical hydrostatic equation using pressure and temperature measured at the ground; thus $\sigma_m(\theta)$ can then be obtained using Rayleigh scattering formulation.

Within this work, the aerosol particles are assumed to be homogeneous spherical particles. The aerosol size distribution was assumed to follow a modified power-law distribution as suggested by McClatchey et al. (1972), given by:

$$\frac{dN_a(r)}{dr} = C_1 \ \text{for } 0.02 < r < 0.01\,\mu\text{m}$$
$$\frac{dN_a(r)}{dr} = C_1\,r\,\Phi \ \text{for } 0.1 < r < 10\,\mu\text{m} \tag{3}$$
$$\frac{dN_a(r)}{dr} = C_1 \ \text{for } r < 0.02\,\mu\text{m and } r > 10\,\mu\text{m}$$

where $N_a(r)/dr$ is the number of aerosol particles with radii between r and $r + dr$ in unit volume, Φ is the size index (or size distribution) and C_1 is the normalization constant that follows from continuity:

$$\int_0^\infty N_a(r)dr = N_{total}\,(\text{cm}^{-3}) \tag{4}$$

17

The total number of aerosol particles, N_{total}, is obtained by integrating $N_a(r)$ over the 0.02–10 μm size range. The aerosol number density profiles have been integrated between 50 and 1100 m altitudes at 12 height intervals to derive ACC (Sharma, 1994). ACC is then considered as aerosol/particulate loading in the boundary layer. The size index and refractive index values of aerosols have been estimated simultaneously by applying the 'inversion by iteration' method to the bi-static lidar angular scattering measurements in the surface layer (Pandithurai et al., 1996) and are utilized in the retrieval method (outlined above) for computing the aerosol number density profiles. Possible errors resulting from these assumptions for determining the size distribution from light scattering data (and hence in the aerosol concentration or column content) have been discussed by Tanaka et al. (1982) and Pandithurai et al. (1996).

Several studies have been carried out to determine the nocturnal boundary layer (NBL) structure and stratification over Pune using the lidar techniques. The multi-year aerosol lidar observations indicate that the height of the NBL at Pune varies from about 200 m to a maximum of 1100 m with a mean contribution from the former to the latter of about 40 % (e.g., Raj and Devara, 1993; Devara et al., 2002). Hence, the integration of aerosol concentration with height was carried out up to the upper limit of 1100 m. Thus the ACC up to 200 m is used to document the contribution of surface-generated aerosols (mostly originated from anthropogenic activities) to the aerosol content in the NBL while the ACC up to 1100 m is used to obtain aerosol content which reasonably describes the aerosol burden in the NBL.

The lidar system was operated every Wednesday and also on alternate Thursday during clear sky conditions. Each aerosol profile obtained from the lidar has been integrated over two different height regimes mentioned. Thus obtained monthly mean values have been utilized in the present study. Additionally, aerosol number density was also obtained at three characteristic altitude levels within and above the atmospheric boundary layer (Devara et al., 2002).

3.3 Radiometer

The solar radiometer employed in the present study is a multi-channel sun photometer that provides on-line data of height-integrated AOD at six wavelengths covering the wavelength band ranging from UV (ultra-violet) to NIR (near infra-red) regions (380, 440, 500, 675, 870 and 1020 nm). There are several methodologies by which AOD measurements can be made. The most common is by measuring the intensity of the direct solar beam and plotting against air mass. The zero intercept of such a Langley plot defines the incoming radiation above the atmosphere, and the slope is the optical depth. If corrections are then applied for Rayleigh scattering and for effects of various trace gases (especially ozone), then the residual is the aerosol optical depth. Here a brief description of the Microtops, which has been used in this study, is reported.

MICROTOPS-II is a multi-channel, hand-held Sun photometer that can be configured to measure total Ozone (O_3), total water vapor or Perceptible Water Content (PWC), and Aerosol Optical Depth (AOD) at different wavelengths. The principal design goal was the measurement of total Ozone to within 1% of ozone measurements made by much larger, heavier and more expensive Dobson and Brewer Spectrophotometers. This instrument has two compact, on-line multi-band solar radiometers (MICROTOPS-II, manufactured by M/s Solar Light Co. USA). A photograph of this instrument is depicted in Figure 3. One of these Radiometers (Sun Photometer) provides height integrated AOD at six wavelengths covering from UV (Ultra-Violet) to NIR (Near Infrared Region) and hence the size distribution of aerosol and the other Radiometer (Ozone and PWC monitor) determines total column Ozone (using the UV band) and PWC (using the NIR band) simultaneously. Both the Radiometers are mounted on a single wooden platform which is in turn fixed to a tripod for achieving high stability, time synchronization between observations and easy focusing of the radiometers to the Sun's disk. An internal microcomputer automatically calculates the site's geographic co-ordinates and universal time, altitude and pressure. The co-ordinates can be altered manually or by a Global Positioning System (GPS) receiver. The main parameters of these radiometers are presented in the Table-1.

Figure 3: MICROTOPS II Sun photometer and Ozonometer used in the experiment

Table-1: Main Characteristics of the Radiometer used in the Experiments

Parameter	Value / Description
Filter Wavelength	**Sun Photometer**
	0.38, 0.44, 0.5, 0.675, 0.87, 1.02 μm
	(0.005- 0.01μm FWHM)
	Ozone Monitor
	0.3055, 0.3125 and 0.32μm
	(0.005μm FWHM)
	Perceptible Water Content Monitor
	0.94 and 1.02 μm
	(0.01μm FWHM)
Field-of-View (FOV)	$< 2.5°$
Dynamic Range	$3×10^4$

Both the radiometers are tailored with optical collimators accurately aligned with their field-of-view and baffles for eliminating the internal reflections. Each channel is fitted with a narrow band filter and a Gallium Phosphide Detector (GaP) suitable for the particular wavelength range. A Sun target and a pointing assembly are permanently attached to the optical block and laser aligned to ensure accurate alignment with the optical channels. The radiation captured, by the collimators and band passed by the filters falls onto the photo-diodes, produces an electrical output proportional to the radiant power (irradiance). These outputs measured at each filter are amplified and analog-to-digital (AD) converted and finally stored, together with the time of observation provided by the built-in master clock in the memory for further analysis. The functional diagrams of sun photometer version of MICROTOPS II are shown in Figure 4.

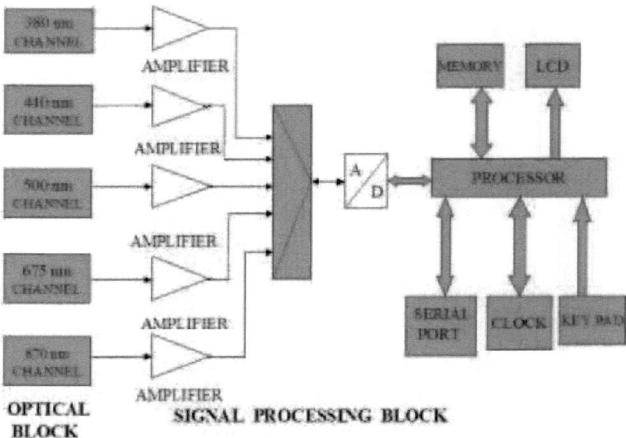

Figure 4: Schematic of sun photometer version of MICROTOPS II

Apart from continuous monitoring of the calibration constant for each channel, inter-instrumental comparison of AOD at either exact or near-synchronous wavelengths has been carried out to ensure the stability and reliability of measured AODs. This instrument has also been operated from the location adjacent to the lidar. Detailed descriptions on both the lidar and radiometer systems have been reported in Devara et al. (1995a; 2001).

The collocated radiometric measurements were performed from morning till evening on the days when the sky was nearly free of visible clouds, and none were near the solar disc and the air mass changes rapidly during the post-sunrise and pre-sunset periods. The total optical depth obtained with each filter was corrected for air molecules (Rayleigh scattering) and also for absorption due to gases around the measured spectral bands in order to determine the AOD at each wavelength.

3.4   Retrieval of AOD

The optical depth of the atmosphere can be determined from the Beer-Bouger-Lambert law, expressing attenuation (a) of the direct solar beam in the atmosphere in the form

$$F(\lambda)=F_0(\lambda)[\exp(-\tau_{total}(\lambda)m(\chi)]$$

where $F(\lambda)$ is the monochromatic solar irradiance reaching the instrument detector at wavelength λ. $F_0(\lambda)$ is the irradiance incident at the top of the atmosphere and it can be used to self-calibrate the system. m is the optical air mass, a function of Solar Zenith Angle (SZA) χ ($m=\sec(\chi)$) for $\chi \leq 75°$). The straight line fitted to the data points of the plot of natural logarithm of radiometer output versus air mass, slope gives the atmospheric total optical depth (τ_{total}), which is given by

$$\tau_{total}(\lambda)=\tau_m(\lambda)+\tau_g(\lambda)+\tau_a(\lambda)$$

where $\tau_m(\lambda)$ is the optical depth due to air molecules Rayleigh scattering; $\tau_g(\lambda)$ is the optical depth due to gas molecular absorption (i.e. due to gases that are expected to contaminate) the measurements, which may be useful in determining the columnar content of the atmospheric species or gases such as ozone, water content and $\tau_a(\lambda)$ is the optical depth due to aerosol particles. $\tau_m(\lambda)$ can be evaluated by knowing the refractive index of air molecules and molecular number density of air at standard atmospheric temperature and pressure. $\tau_g(\lambda)$ can be computed from absorption cross section of different gas molecules at any wavelength and from known number density of the absorbing gas. $\tau_a(\lambda)$ can be computed by subtracting $\tau_m(\lambda)$ and $\tau_g(\lambda)$ from $\tau_{total}(\lambda)$.

The radiometers are operated initially by keeping the cover closed for the optical blocks (consisting of windows, filters etc.). During this period the instrument stores the background values for all the filters. In the next few seconds, on removing the cover it collects a set of over 25 observations for each filter. The average value thus obtained for each filter, is used to compute the spectral variation of columnar AOD, ozone and perceptible water content instantaneously, and are depicted on the display for a quick look, and stored in the memory. Ozone absorbs shorter wavelengths of UV radiation much more than longer wavelengths. This means that the amount of ozone between the observer and the Sun is proportional to the ratio of two wavelengths of the Sun's UV radiation. The ozone monitor used in the present experiment utilizes this relationship to derive total ozone column (the equivalent thickness of pure ozone layer at STP) from the measurements of two wavelengths in the UV region (0.3055 & 0.320 μm). The measurements at additional third wavelength (0.3125 μm) enable correction for particulate scattering and stray light. The columnar PWC is determined based on the measurements at 0.940μm (H_2O absorption peak) and at 1.020μm (no absorption by water content). The AOD at 1.020μm is also calculated based on the extra-terrestrial radiation at that wavelength, corrected for sun-earth distance, and the ground level measurements of the radiation at 1.020μm (Devara et al., 2001; Morys et al., 2001).

3.5 Observations and analysis

The AOD (at 380, 440, 500, 675, 870 and 1020 nm), O_3 and PWC data sets used in the study were from May 1998 to October 2003. Care was taken to collect the observations only when the sky was nearly free from clouds and none were near the line-of-sight to the sun. The observations were repeated at close intervals between 10-15 minutes during the post-sunrise and pre-sunset period when the air mass changes rapidly, 30 minutes interval in the intervening period. Thus the data presented in this report corresponds to the measurements made during cloud-free conditions only and as such the present observations are free from hydrometeors whose sizes are far larger and less efficient in their interaction with light at

24

wavelengths considered here. A total of more than 20 sets of observations of direct solar radiation were obtained on each observational day. For brevity, AOD at three characteristic wavelengths (380, 500, 1020 nm) only have been used in the Wavelet Transform analysis. The main aim here is to get the small, sub-micron and coarse mode particle variations corresponding to the wavelengths 380, 500 and 1020 nm respectively. For the diurnal variation, data collected on very stable days recorded at all the six (380, 440, 500, 675, 870 and 1020 nm) wavelengths have been considered. The data could not be obtained on some experimental days either part or full day due to cloud hindrance. During such occasions, multiple regression analysis techniques have been used to obtain the AOD, O_3 and PWC for the data gaps, particularly during the monsoon months (June, July, August, and September).

4. Wavelet transforms analysis

Wavelet transform analysis is an advanced statistical tool for analyzing localized variations of power within a time series. By decomposing a time series into time/frequency space, one is able to determine both the dominant modes of variability and how those modes vary in time. The WT can be used to analyze time series that contain non-stationary power at different frequencies (Daubechies, 1990). In wavelet-based spectral analysis, a generalized local base function (mother wavelet) is dilated and translated with flexible resolutions in both frequency and time so that a family of wavelets (daughters) is generated. Finally, these daughter wavelets are expressed as function of scale (s) or period and time position or translation (n). In general, when a time-series signal of interest is convoluted with some daughter wavelet, a wavelet coefficient is obtained for a specific point in the time frequency domain. Such analysis is repeated for different combinations of n and s (i.e. other daughter wavelets) which yields a set of wavelet transform coefficients representing the decomposition of the signal into time-frequency space. The unique feature of the WT analysis is that it can decompose a given time series into a multi-resolution series providing an insight into the likely causative influences that operate at various scales (large-scale as well as small-scale localized processes). The WT thus allows the wavelet to be scaled to match most of the high- and low-frequency signals so as to achieve the optimal resolution with the least number of wave functions. This 'zoom-in' property is an extra-ordinary feature of WT.

Different steps involved in the wavelet-based spectral analysis have been reported in the past by many researchers (e.g., Torrence and Compo, 1998) so that only a brief overview on the algorithm applied is presented. Some standard and frequently used wavelets are the Haar wavelet (step function), the Marr wavelet (second derivative of the Gaussian), the Morlet wavelet, or the Daubechies wavelets (Daubechies et al., 1992; Torrence and Compo, 1998). An important step at the beginning of the WT analysis lies in the choice of a suitable wavelet function. Within this work, wavelet spectral analysis has been applied to identify correctly different modes of variability or periodicities embedded in the long-term time series of lidar-derived ACC and radiometer-retrieved spectral AODs; Morlet-based wavelet analysis is an appropriate choice for this (Torrence and Compo, 1998). The classical Morlet mother-wavelet is basically formed by a plane wave modulated by a Gaussian function which is defined as

$$\Psi_0(\eta) = \pi^{-1/4} e^{i\omega_0\eta} e^{-\eta^2/2}, \tag{5}$$

where η is a non-dimensional time parameter and ω_0 is the non-dimensional frequency parameter, here considered to be 6 to allow admissibility condition of the function (Farge,

1992). Once a mother wavelet function is selected, it is necessary to choose a set of scales 'S' to use in the WT. It is convenient to write the scales as fractional power of two as

$$S_j = S_0 \, 2^{j\delta},$$

(6)

where S_0 is the smallest resolvable scale and $j = 0,1,2.....J$. J determines the largest scale. S_0 is chosen in such a way that the Fourier period is approximately equivalent to 2 δt where δt is the temporal resolution. The choice of sufficiently small δj depends on the width in the spectral space of the wavelet function. It is important to note that the Morlet wavelet yields l = 1.03 s, where l is the Fourier period, indicating that for the Morlet, the wavelet scale s is almost equal to the Fourier period (Torrence and Compo, 1998). The terms 'scale' and 'period' will be used synonymously henceforth.

The CWT $W_n(s)$ of a discrete time series $x_{n'}$ (here time series of lidar or radiometer derived aerosol parameters with time resolution of δt and N number of points or samples) at time t_n = n δt for a scale s is defined as the convolution of x_n with a scaled and translated version of mother wavelet $\Psi_0(\eta)$:

$$W_n(s) = \sum_{n'=0}^{N-1} x_{n'} \Psi^* \left[\frac{(n'-n)\delta t}{s} \right],$$

(7)

where the (*) indicates the complex conjugate. In Eq. (7), n and s govern respectively the translation of the function along the time coordinate and the scales represented by the function. By varying the wavelet scale s and translating along the localized time index n, the wavelet power spectrum is obtained.

Wavelet function (Eq. (5)) is in general complex; consequently, the wavelet transform $W_n(s)$ is also complex and consists of an imaginary part and a real part or an amplitude and corresponding phase. The wavelet power spectrum is then defined as the absolute value squared of the wavelet transform which is denoted by $|W_n(s)|^2$. It should be noted that it is possible to calculate the wavelet transform using Eq. (7). But, the convolution in Eq. (7) is implemented in Fourier space, which is considerably fast (Torrence and Compo, 1998).

To achieve CWT, the convolution in Eq. (7) should be performed N (number of points in the time series) times for each scale s. To ease the numerous numbers of convolutions to be performed simultaneously, discrete Fourier transform (DFT) is applied. The DFT of x_n is defined as

$$\hat{x}_k = \frac{1}{N}\sum_{n=0}^{N-1} x_n \, e^{-2\pi i k n / N} \, , \tag{8}$$

where $k = 0......N - 1$ is the frequency index. On applying the convolution theorem, one can derive the wavelet transform as the inverse Fourier transform of the product:

$$W_n(s) = \sum_{k=0}^{N-1} \hat{x}_k \, \hat{\Psi}^*(s\omega_k) e^{i\omega_k n \delta t} \, , \tag{9}$$

where the angular frequency is defined as

$$\omega_k = \frac{2\pi k}{N \, \delta t} \text{ for } k \le \frac{N}{2} \, , \text{ and} \tag{10a}$$

$$\omega_k = -\frac{2\pi k}{N \, \delta t} \text{ for } k > \frac{N}{2} \tag{10b}$$

Thus, by using Eq. (9) and a standard FFT algorithm, one can calculate efficiently the CWT. Herewith readers are referred to the comprehensive description of the CWT by Torrence and Compo (1998) for obtaining a further detailed overview on the normalization, choice of scales, re-construction of time series from the spectra, etc.

The CWT finally yields the local wavelet power spectrum $|W_n(s)|^2$ so that a representation of time versus scale/period cross-section (often referred as wavelet scalogram) of the amplitude

28

variability is obtained. In a scalogram, $|W_n(s)|^2$ are represented with a predefined pseudo-color scale. The bottom-axis is then time and vertical-axis refers to the scale or period. Thus, the wavelet energy scalogram helps to figure out how the amplitude varies in time. Thus, if a vertical slice on the wavelet scalogram is considered, a measure of local wavelet spectrum can then be obtained. Similarly, one can also obtain a time-averaged (slice) wavelet spectrum over a certain period. Such calculations can be performed for each time step to generate a wavelet plot smoothed by a certain window (Torrence and Compo, 1998). If an averaging is performed over all the local wavelet spectra, one can obtain then the corresponding global wavelet spectrum

$$\overline{W}^2(s) = \frac{1}{N} \sum_{n=0}^{N-1} |W_n(s)|^2 \ .$$

(11)

The global wavelet spectrum provides an unbiased and consistent estimation of the true power spectrum of a time series. It can also be used as a measure of background spectrum against which peaks with different periodicities can be examined.

To explain the different periodic features obtained in the scalogram, it is necessary to perform a detailed statistical significance test. Following Torrence and Compo (1998), this test is done for local wavelet spectra by choosing white noise (with a flat Fourier spectrum) as the background spectrum. A chi-square test is performed to determine the 95% confidence (significant at 5 %) that the wavelet will fall in a specific range of scales and times. Period or frequencies with the power less than the 95% Fourier confidence spectrum (95 % line) are then considered statistically significant.

With an aim to resolve the dominant periodicities, the Morlet-based CWT analysis was performed on the original (non-smoothed) time series of lidar-derived ACC as well as radiometer-retrieved AOD obtained over Pune. The potential of CWT-based analysis, in particular, the time-frequency localization, helped delineate signatures of long-period oscillations and their evolution in time.

5.   Results and Discussion

5.1   Diurnal Variation of Aerosol Optical Depth (AOD), Total Column Ozone (TCO) and Perceptible Water Content (PWC)

Variations in columnar AOD at 6 wavelengths (380nm, 440nm, 500nm, 675nm, 870nm, 1020nm) on four typical clear-sky days during different times of the day are plotted in Figure 5. Also the TCO and PWC variation throughout the day is also plotted in Figure 6. It is evident from the frame1of the figure that the time variation of AOD follows the spectral dependence i.e. higher AOD are observed at shorter wavelengths and vice-versa, except at 380nm where AOD showed smaller values as compared to its next longer wavelength i.e. 440nm. This event suggests conversion of lower size range of accumulation-mode particles into sub-micron size particles due to gas-to-particle conversion processes.

All the graphs show that the AOD attains the high value at the afternoon hours and low value during the morning and evening hours for all the wavelengths. Convective turbulence processes (In cloud-free air) cause mixing of particles and lifting of fresh lighter aerosols particles that are generated due to anthropogenic activities around the experimental site. Besides the changes in AOD during afternoon hours could be explained on the basis of advection of pollutants from surroundings and convective activity leading to changes in aerosols particle number distribution and gas-to-particle (photochemical processes) while those during forenoon hours could be attributed to the impacts of radiative cooling and turbulent processes and aerosol characteristics during previous night. Break up of inversions and associated ventilation of aerosols particles and their further modifications during afternoon hours may cause significant differences in the AODs during the forenoon and afternoon hours. The day-to-day variations observed in the AOD over and above the basic variation, could be due to changes in local meteorological parameters that induce mixing of aerosols particles of different sizes and chemical compositions in different regions of the atmosphere. A peak is observed in the morning almost on all days due to local traffic.

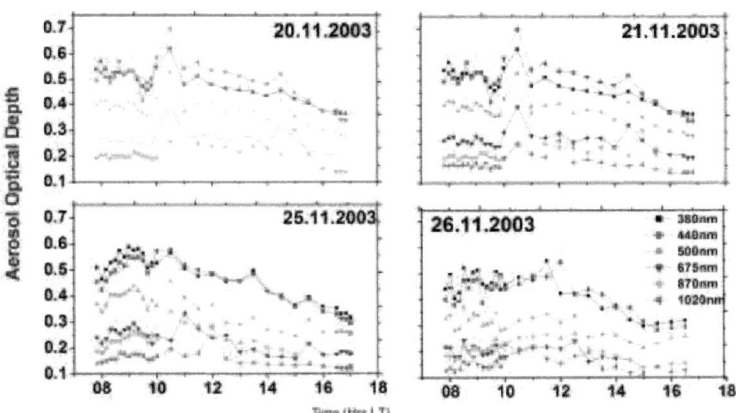

Figure 5: Daytime diurnal variation on AOD during four typical clear sky days

Figure 6 shows the daytime diurnal variations of TCO and PWC for the above four days. The TCO shows in all four days two peaks, one during the morning time and the other during the afternoon hours. These results suggest that the ozone amounts depend upon the nature and variations of the anthropogenic activity, particularly during peak traffic hours in and around the experimental station.

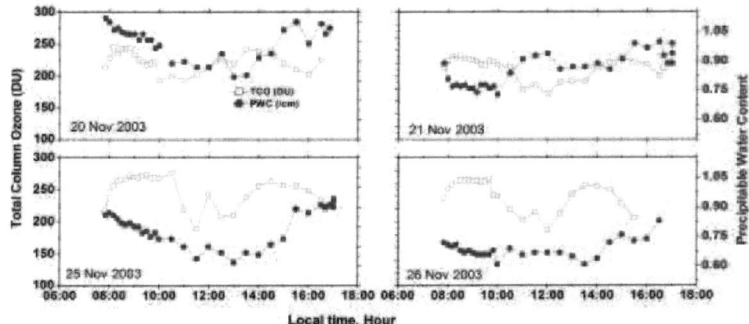

Figure 6: The diurnal variation of TCO and PWC

The first peak may be attributed to the anthropogenic activities and the second one due to the photochemical reactions in the afternoon hours. The PWC shows in the entire figure a common feature that it is in increasing trend from morning to the evening hours with a broad minimum during afternoon hours.

5.2   Wavelength dependence of AOD

The spectral optical depth, a function of the number, size, composition and shape of the particles present in a given column of air, is a fundamental optical property of the atmosphere. Sun-photometry has been proven to be a successful technique for measuring multi-wavelength AOD (Yamamoto & Tanaka, 1969; King et al., 1978; Devara et al., 1996). Figure 7 portrays the spectral distribution of AOD observed on four clear sky days in the month of November 2003. On close inspection, these plots reveal that AOD decreases with increase in wavelength, which is in accordance with the Mie-Theory. In other words this feature is consistent because the sub-micron particle concentration always dominates the concentration of larger particles in the atmosphere.

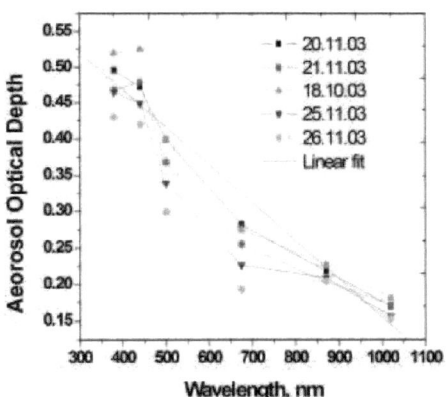

Figure 7: Spectral dependence of the AOD. Navy line indicates the linear fit to the data

5.3 Long-term changes and trends in aerosol distribution

The long-term measurements of ACC between October 1986 and September 2000, and radiometer-retrieved AOD between May 1998 and December 2003 have been first subjected to the regression analysis to examine the long-term changes and trends in the aerosol loading over the experimental station. Additionally, the ACC up to 1100 m along with aerosol number density at three characteristic altitudes of 50, 1000 and 1800 m, and the radiometer-measured AODs at three typical wavelengths of 380, 500 and 1020 nm, representing fine, sub-micron and coarse-mode aerosol particles, respectively, have been used for the CWT analysis. For obtaining the wavelet spectra, monthly mean values of the above mentioned parameters are used here to smooth out the short-term and mesoscale fluctuations in the ACC and AOD data sets. Thus, the total number of elements in the time series remains either 168 (for lidar) or 66 (for radiometer).

The aerosols present in the lowest air layers, particularly within the ABL, typically up to 1100 m in the present study (Raj and Devara, 1993), contribute greatly to the overall loading at the experimental site, and thus this information is of paramount importance. Before presenting the multi-scale features in the aerosol observations associated with the different oscillatory behavior of the aerosol loadings over the station, some general statistical information on the ACC and AOD regimes is given. Figure 8 shows the monthly mean aerosol content, derived from the vertical distributions of aerosol number density measured over Pune during 14 years period. Additionally, annual mean values for each year are also presented which further confirm an increasing trend in the ACC. It is evident from the figure that the aerosol loading over the station exhibits an increasing trend, which is in agreement with the observation of the growth in urbanization, industrialization and land-use pattern changes in the proximity of the station during these years as will be shown in Section 5.5. These variations in aerosol loading and hence in the trend are found to have close relationship with the aerosol generation and removal processes, particularly the washout phenomenon during the south-west monsoon season, prevailing over Pune (Devara et al., 2003).

Seasonal variability of the optical properties of aerosol particles is governed mainly by the circulation patterns and aerosol source of origin. The experimental site including its

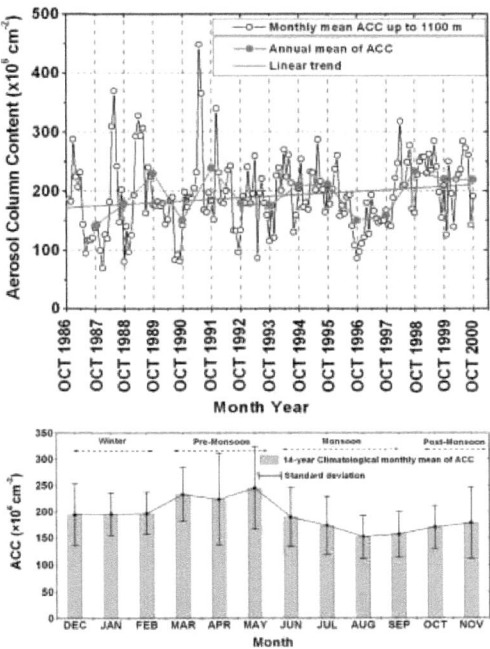

Figure 8. Monthly mean variations in ACC up to 1100 m (open circles) between 1986 and 2000. The dotted line represents the long-term trend. Annual mean values (closed circles) are also shown (upper panel). Climatologically monthly mean values of observed ACC for the study period are shown in the lower panel. The error bars on ACC show the standard deviation (implying the variability) within a month during different years.

surrounding area experiences tropical weather with four clearly distinctive seasons: pre-monsoon (March–May), southwest monsoon (June–September), post-monsoon (October–November) and winter (December–February). To demonstrate the month-to-month development in the observed ACC, climatological means (global means for the entire period from 1986 to 2000) for each month have been computed as shown in Figure 8 (lower panel).

The vertically-aligned error bars on the ACC columns mark the standard deviation obtained while computing the mean monthly values, thus, correspond to the variability of ACC for each month over entire observation period.

Figure 8 clearly yields that the pre-monsoon season is not only characterized by the higher ACC over all other seasons but also by higher variability (see, standard deviations overlaid on the figure) pertaining to the predominant convective features and frequently occurring dust storm activities during this hot season which most probably triggers the accumulation of aerosol in the boundary layer. The entire cycle of the lidar-derived monthly mean ACC values clearly evinces that the aerosol loadings within the boundary layer over Pune is minimum during the southwest monsoon months (often referred to as summer monsoon), gradually increases during the post-monsoon and winter, and finally attains highest values during the pre-monsoon months. While comparing the corresponding seasonal mean values of ACC, aerosol burden during the pre-monsoon season (global mean of 234.09×10^6 cm^{-2}) over the site is found to be higher by 39.25 % than that during monsoon season (168.12×10^6 cm^{-2}) and by 19.61 % and 34.25 % than that during winter and post-monsoon, respectively.

The collocated solar radiometer derived AODs, averaged from respective months during 1998-2003 for three characteristic wavelengths, 380, 500 and 1020 nm are shown in Figure 9. The monthly mean AODs show well-defined wavelength dependence i.e. smaller AOD at longer wavelength and vice-versa, as expected according to Mie theory (Pinnick et al., 1973). Additionally, the AOD variations at all three wavelengths considered are found to exhibit an increasing trend. Unfavorable weather conditions such as heavy precipitation and frequently occurring optically thick clouds over the experimental station during the south-

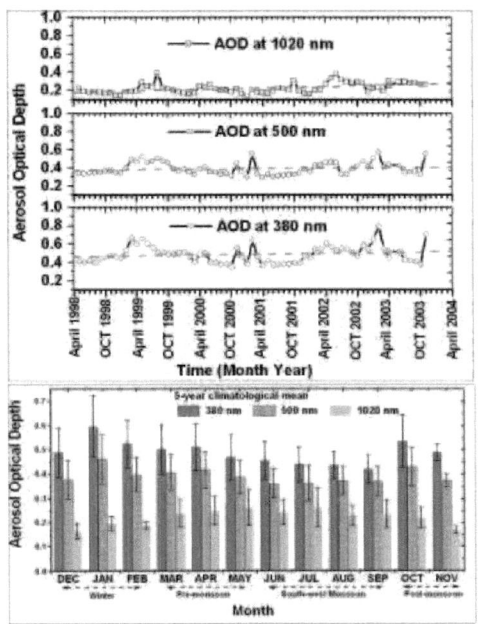

Figure 9. Upper panel: Long-term variations in radiometer-retrieved AODs obtained at 380 (top), 500 (middle) and 1020 nm (bottom) during the period between April 1998 and October 2003. The dashed lines denote the long-term trends embedded in the corresponding time series. Lower panel: Variability in the monthly mean AODs at three wavelengths averaged over 5 years (lower panel). The error bars on the AODs show the standard deviation estimated implicating the variability within a particular month of different years.

west monsoon season hampered the measurements, which yielded gaps (a few days to one week) in the time series. Such gaps in the data set were bridged by the values obtained by applying polynomial regression analysis to the time series. However, it should be noted that instrumental problems were not encountered; thus no spurious values appeared in the retrieved time series. Moreover, while performing the sensitivity tests, this analysis technique was found to reduce the RMS error considerably.

The time series analysis of the long-time measurements reveals that high AOD values at all wavelengths are dominated around the year 1999 while during 2002-2003 similar highs were also present but only for AODs at 380 and 500 nm. Several researchers reported similar high AODs over India in 1999 while examining the measurements performed during Indian Ocean experiment (see, Ramanathan et al., 2001; Moorthy et al., 2001; Devara et al., 2001; and references therein). Following an overall statistics of the two long-term time series, the global mean of ACC value (over 14 years) is found to be 192.55 ±60 ($\times 10^6$ cm^{-2}) while the overall increase was 34.29 ($\times 10^6$ cm^{-2}) with an annual increasing rate of ~2.5 ($\times 10^6$ cm^{-2}). Similar analysis performed for the AODs revealed that the overall increase in AOD at 1020 nm was higher (0.139) than that for other two wavelengths with an increasing rate of 0.019 yr^{-1}.

Besides the wavelength dependence, the long-term (increasing) trend in the AOD at 1020 nm is found to be prominent as compared to those at 380 nm (0.009 yr^{-1}) and 500 nm (0.005 yr^{-1}). This could be attributed to the long-range transport of aerosols of marine origin which occurs often during the transition period between the pre-monsoon and monsoon seasons over the station. In order to identify the long-range transport, analytical backward trajectories were used (not shown here) available on the AERONET (AErosol RObotic NETwork) website. These trajectories are mainly based on a kinematic trajectory analysis using NASA GMAO (Global Modeling Assimilation Office) assimilated gridded analysis data. In general, backward trajectories provide some key information about the possible origin of the observed aerosols and about the synoptic patterns corresponding to the measurements. 7-day back trajectories for air-parcels arriving over Pune (an AERONET site) at an altitude of 1500 m (850 hPa) for the arrival time of 12:00 UTC were used for this purpose. These analyses reveal that the south Indian Ocean and the adjacent Arabian Sea are the main sources of the marine aerosols transported over the site. Transport of coarse-mode particles over the experimental station due to strong winds and convective activity during the pre-monsoon and monsoon months cannot be ruled out. This is conceived that more detailed trend analysis for further investigation of the possible cause of the observed increasing trend in the aerosol load over the site with the aid of air-mass provenance back-trajectory calculations (e.g., by cluster algorithm) and associated meteorological parameters is beyond the scope of this work. However, these analyses will definitely help to single out the role of long-range transport and synoptic meteorology in modulating the aerosol variability over the site.

Additionally, climatological monthly mean values of AODs at three wavelengths to investigate the impact of the different seasons dominating around the site have been studied. Intra-seasonal variations including month-to-month variability in the spectral AODs for the entire time period from 1998-2003 are shown in Figure 9 (lower panel). On every year from 1998 to 2003, monthly means of AOD show gradual increase of aerosol loading from December to April. The error bars display the corresponding standard deviation which serves as an indicator of inter-annual variability. It can be seen that from winter to pre-monsoon season, the AODs at 380 nm and 500 nm slowly decreases. Interestingly, an exactly opposite seasonal characteristic (an increase from winter to pre-monsoon) can be observed for AOD at 1020 nm. After south-west monsoon passes the station typically end of September, AODs at smaller wavelength (mostly originated due to local anthropogenic activities) continues to increase slowly till the end of winter: seasonal mean of AODs during winter is higher by 23 % (380 nm) and 12 % (500 nm) than in monsoon season. In contrast, AODs at 1020 nm starts to decrease during winter. This could be attributed to the stable boundary layer regime associated with weak surface winds prevailing often during the winter months over the site which favors smaller number of larger sized aerosol particles (equivalent to AOD at 1020 nm, e.g., soil dust) during this season. But, during pre-monsoon season, relatively higher abundance of larger sized particles around the site most probably yields the high values of AODs at 1020 nm. Seasonal mean AOD at 1020 nm for pre-monsoon season is found to be 36 % than in winter months. Another striking feature which is easily visible is that during pre-monsoon season in 2002 higher values of AODs for all wavelengths are observed when compared to those during the pre-monsoon in 2001. Devara and Raj (1997) also addressed in detail the effect of two contrasting monsoon seasons (1987-1988) on the lidar derived ACC distributions. The relationship between aerosol concentration/extinction/optical depth and south-west monsoon activity has been extensively studied (Devara et al., 2003).

Finally, inter-annual variability in the aerosol distribution obtained with the lidar and radiometer measurements at the experimental is shown in Figure 10. It can be clearly seen that the spectral AOD at 380 nm undergoes a significant year-to-year variability from 1998 to 2003 including the dominant annual variability. Corresponding annual means of AOD also show year-to-year variability. It is interesting to note the remarkably high AOD in 1999. Li and Ramanathan (2002) also reported high values of AOD in 1999 as compared with 1996 and 1998 over different parts around India.

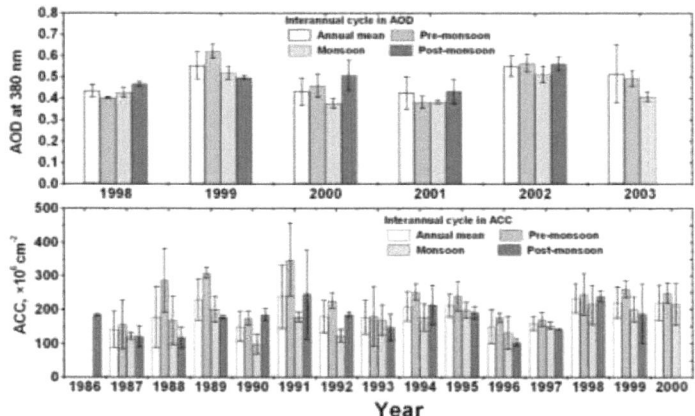

Figure 10. Interannual variations of radiometer-retrieved AOD at 380 nm (upper panel) and lidar-derived ACC (lower panel) during the pre-monsoon, summer monsoon and post-monsoon seasons over the experimental site. Yearly mean values are also shown. The error bars display the plus or minus one standard deviation in ACC and AOD indicating the variability.

A strong inter-annual variability in the lidar-derived ACC can also be noticed from Figure 10. Relative standard deviation of year-to-year variability in seasonal mean AOD and ACC were estimated which yielded that the pre-monsoon AOD and ACC undergo maximum variability of around 20 % during 1998-2003 and 1986-2000. On the other hand, variability in the annual mean AOD and ACC were 12 % and 18 %, respectively.

5.4 Wavelet analysis of the lidar-measured aerosol distribution

Time series of the monthly mean values of the ACC is first used to study the variation of power spectrum with respect to the scale or period in the entire time domain mentioned.

Figure 11 (upper panel) shows the pseudo-colored scalogram (time/period) of the normalized wavelet power spectrum (average of absolute wavelength coefficient squared) for the aerosol loading up to 1100 m above ground level (AGL). In this scalogram, localized wavelet power spectra are represented with pseudo-color scale increasing from black through green to red as shown by the color bar. The X-axis in Figure 11 (upper panel) is time while the Y-axis refers to the scale or period ranging from 2 to 32 months. Contours are also over plotted on the colored amplitudes in the scalogram.

Localized wavelet spectrum in the scalogram indicates wave signature with a period of around 20 months. The global wavelet spectrum also shows a peak around this period/frequency (Figure 11, lower panel). It is clear from the figure that most of the power is concentrated within the bands of 12-14 months and 20-22 months, which are considered to be annual oscillations (AO) and quasi-biennial oscillations (QBO), respectively. Some signatures of 28-month periodicity can also be observed, however, its strength is lower compared to the other dominant periodicities. It should be noted that exact period (in months here) of QBO is probably not well-defined. Reed et al. (1961) found a variable period of QBO in the wind field ranging from 21-32 months while Kane (1992) reported periodicities (of QBO) in the concentration of trace elements and surface aerosols in the range of 2-3 years. The AO is observed to be present throughout the study period with varying intensity. The CWT analysis also determined the global wavelet spectra which are nearly equivalent to the power spectra from an FFT (see, lower panel in Figure 11). The scales (periods) corresponding to a value of around 20-22 months with higher value of wavelet coefficient (red-color in the scalogram) significantly demonstrates the presence of the corresponding frequency between 1990 and 1991. The results concerning the significance level test show a coherency with the global wavelet spectrum for the ACC in the air layer of 1100 m. It is evident that both periodicities related to the AO and the QBO are dominant and are well matched with the peaks exhibited by the corresponding global wavelet spectrum. The periodicity around 22.5 month (QBO) is found to be more significant than the AO. However, the strength of the observed period of QBO in ACC decreases at the end of the time series possibly due to the local modification of the aerosol distributions and changes in the ABL depth.

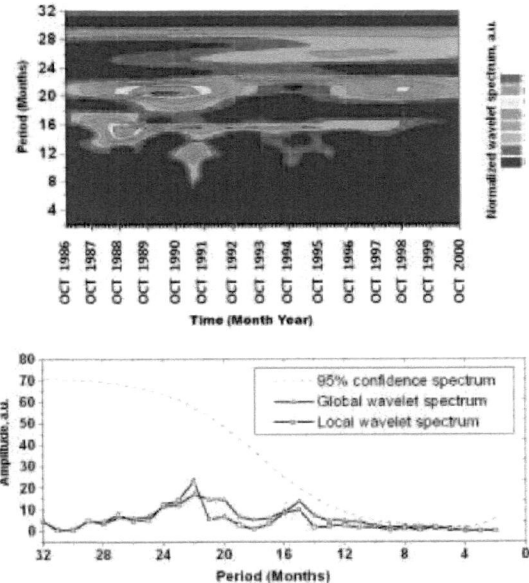

Figure 11. Upper panel: wavelet scalogram of the time series of lidar-derived ACC up to 1100 m AGL (lower panel). In the scalogram (time/period), the amplitude of the normalized wavelet power spectrum is color coded increasing from blue to red. Lower panel: comparison among the global wavelet spectrum (equivalent to Fourier spectrum), local wavelet spectrum, and 95% confidence spectrum (dashed-line) for the purpose of significance level test. See Section 5.2 for further discussion.

5.5 Wavelet analysis of aerosol concentration retrieved for different altitudes

In order to better characterize the periodicities present in the time series of ACC, wavelet-based spectral analysis was further applied to the ACC obtained at three different altitudes. From the weekly vertical profiles of aerosol concentration, monthly mean concentration values at three characteristic heights (50, 1000 and 1800 m AGL) were computed and subjected to WT analysis to examine the presence and propagation of different dominant periodicities present in the long-term measurements. Thus, the time variations of the corresponding CWT spectra helped to examine whether the periodicities (waves) present at each altitude are evanescent or propagating type.

The CWT spectra calculated for the aerosol concentration observed at three selected heights are shown in Figure 12. Corresponding global spectra were also calculated (not shown) for confirming the presence of the periodicities present in the scalogram. The presence of three characteristic periodicities of around 12, 16 and 23 months is evident from the scalogram for 50 m AGL. The important feature which can be evidenced from the figure is that the wave with periodicity around 12-13 months (close to AO) has the largest amplitude as compared to that observed for the periodicity around 24 months (close to QBO). Since the aerosol concentration observed at 50 m AGL is highly influenced by local changes of the near-surface aerosol content perturbed by the local pollution events, remarkable variations from synoptic (about one week) scale is expected. But, monthly resolved data set are used so that the weekly periodicities have not been extracted from the CWT analysis. However, with the results presented, it is possible to characterize both AO and QBO in detail. For instance, there is a significant increase in the intensity of the 16 months periodicity observed during 1992 and thereafter as compared to that around 1988.

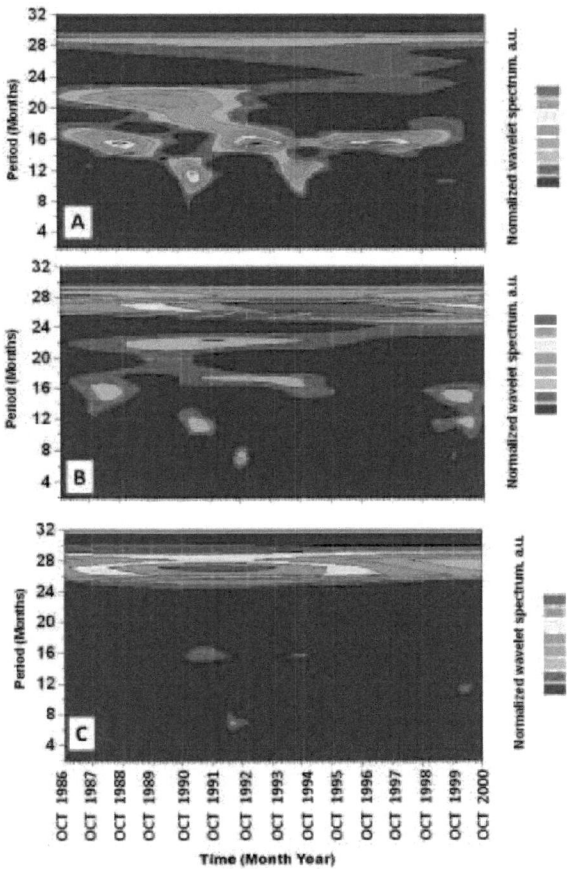

Figure 12. Wavelet scalograms for aerosol concentration calculated for three different altitudes: Top panel is for 50 m AGL, middle panel is for 1000 m AGL, and bottom panel is for 1800 m AGL.

The middle panel in Figure 12 displays the similar spectra calculated for the time series of ACC at 1000 m level. Although the presence of basic periodicities remain same, the dominance (in both strength and significance) of periodicity around 27 months for the ACC at 1000 m AGL is quite clear while comparing the wavelet spectra for ACC at 50 m AGL. Better agreement between this periodicity and global wave spectrum provides further supports to its presence in the time series.

The scalogram representing the time variability of the ACC observed at an altitude of 1800 m AGL (bottom panel in Figure 12) clearly shows that unlike the periodicities observed at 50 and 1000 m AGL, the QBO at 1800 m AGL seems to be stronger and significant with respect to power. The global wavelet spectrum (not shown here) further confirms this aspect. A definite structure of 27 months periodicity with a maximum during the year 1991 can be seen in the figure. Another important feature is that the QBO is much stronger as compared to the AO and semiannual oscillations. This aspect is quite interesting and being investigated in detail separately by analyzing the solar-geophysical and other *in-situ* data. These results suggest that the CWT analysis is an efficient tool to provide information on the evolution (in time) of the different scales or periods associated with the wave-like phenomena that are present in a long-term data sets.

The CWT application on the three different time series shows that the wavelet decomposed periods were present in all three altitudes but at different times which indicates propagation of the waves. For instance, by comparing the three wavelet spectra and their corresponding global spectra, it can be concluded that the longer period waves that are generated from the variations in the aerosol concentration propagate from lower to higher altitudes. This implies that the contribution to aerosols over the experimental site comes primarily from long-range transport (most probably due to wind-blown dust) involving varying air-mass rather than from local anthropogenic sources, which are mainly confined to the surface layer. This is an important aspect which could be furthered by analyzing simultaneously measured vertical distributions of meteorological parameters or simulated back trajectories of air-masses at different levels over the station. No similar studies dedicated in the past to the wavelet applications on the long-term ACC data set are available in the literature so that a comparison of the results discussed was not possible.

It is well known that the variations in local meteorological parameters such as wind, temperature, visibility and so on modulate the time variations in aerosol concentration (Devara et al., 1994) and hence the resultant oscillations propagate both vertically and horizontally through large-scale circulation. Tropospheric aerosols generally show annual and seasonal variability (Bodhaine et al., 1981). In addition to these characteristic oscillations in aerosols, Dutton (1992) reported a strong association between AO in atmospheric transmission factor (ATF) and QBO, and pointed out that the most likely candidate for the source of the QBO in the ATF time series is tropospheric aerosols of the size range 0.1 and 1 μm that scatters solar radiation.

In the wavelet scalogram of ACC at different heights, additional signatures having periodicities more than 28 months can be noted. However, the significance of these structures is questionable due to edge effects of the WT-generated spectra. Edge effects arise due to finite length of the time series under study. In the present analysis, the time series is padded at the end with sufficient zeroes to bring the total length N (total number of samples) up to the next-higher power of two which help to speed up the Fourier transformation during calculation of CWT coefficients (see, Section 4 and Torrence and Compo, 1998 for complete details on the edge effect). Padding with zeroes introduces discontinuities at the endpoints and as one goes to larger scales, the amplitude decreases near the edges as more zeroes enter the analysis; therefore, any kind of interpretation of the results could lead to erroneous conclusions, so has been avoided.

5.6 Wavelet spectra of the variability in the aerosol optical depth

Wavelet spectra as determined from the time series of the radiometer-retrieved AOD measurements obtained at three wavelengths for the period of more than five years are shown in Figure 13. It is clear that the periodicity of around 22 months (close to QBO) is predominant during 2000-2001. Although the AO is also present, its amplitude is very weak as compared to that of the QBO. This feature has also been confirmed by the corresponding global wavelet spectra as displayed in Figure 13.

The spectra obtained for the time series of AOD at 500 nm as shown on the middle panel in Figure 13 exhibits dominance of 23 months periodicity and AO but during 2002-2003. The additional feature of semiannual oscillation with reasonable amplitude during 2001-2002 is consistent as confirmed with the significance level spectrum (Figure 14). While analyzing the spectra for the AOD at 1020 nm wavelength (lower panel in Figure 13), it can be observed that unlike the case of 380 and 500 nm, this exhibited dominant periodicity around 14 to 16 months (AO) during 2002-2003 while the QBO is also found to be present with lower strength. By comparing the global wavelet spectra (Figure 14) corresponding to AOD, representing fine (380 nm), sub-micron (500 nm) and coarse-mode (1020 nm) particles, it is found that the wave with periodicity around 22-23 months has the maximum power and most significant of all the waves present in the time series. The interesting result is that the amplitude of the QBO increases monotonically with AOD at decreasing wavelength. This clearly suggests that the coarse mode particles do not play any significant and governing

role in the processes giving rise to QBO in AOD. Some important and key features of the

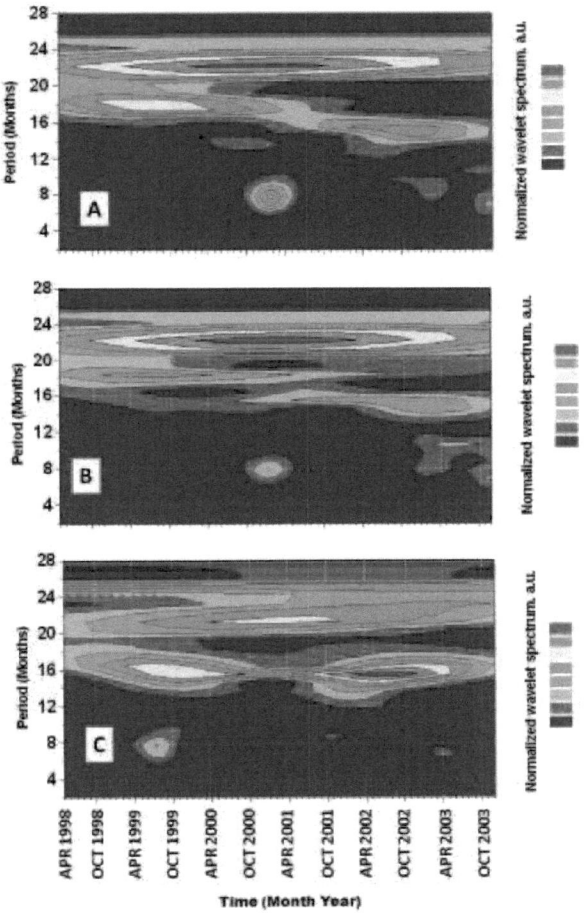

Figure 13. Wavelet spectra for time series of AODs presented in Figure 11. A: AOD at 380 nm, B: AOD at

500 nm, and C: AOD at 1020 nm. See Section 5.4 for further details.

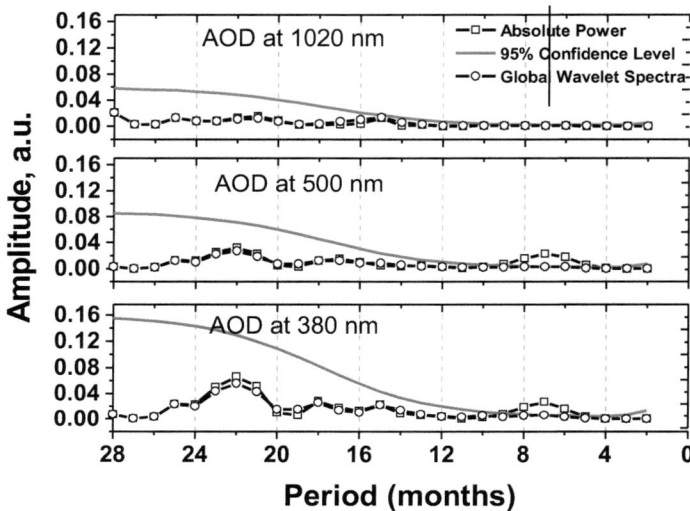

Figure 14. Spectrum of different periodicities, their level of significance and global wavelet spectra matching for AOD at 1020 nm (upper), 500 nm (middle), and 380 (lower) nm.

QBO in AOD at visible wavelength were delineated in their work. However, they did not show any detailed and conclusive results on the relationship between the strength of the QBO in AODs and the spectral wavelengths ranging from UV to IR region as has been identified here.

On the other hand, 14-16 months periodicities are more prominent for AOD at 1020 nm compared to that at two other wavelengths. The 8-months oscillations did not appear prominently in the scalogram for AOD at 1020 nm. But, during the period between 2000 and 2001, this scale is clearly observed in the scalogram for AODs at 380 and 500 nm. It should be noted that these two wavelengths lie in the spectral region where the fine or submicron aerosol particles in their accumulation regime contribute to the AOD. These particles are believed to be mainly originated from secondary production mechanisms like gas-to-particle conversion processes. Thus, they are highly susceptible to anthropogenic activities (Ramanathan et al., 2001). Unlike the coarse-mode particles which are in general short-lived, the fine particles have a larger residence time in the atmosphere and they are influenced by long-range transport episodes. AODs at shorter wavelengths are therefore considered to yield effects that are more local in nature. This clearly indicates that the particles in the accumulation mode significantly contribute to the QBO in AODs as have been evidenced here.

5.7 Influence of urbanization characteristics on aerosol load

Possible contribution of the urbanization on the aerosol burden over the station by studying the co-variations between urbanization parameters and lidar derived ACC estimates is aimed. Pune Municipal Corporation (PMC, Pune) provided the key data set on the city's population distribution, population growth, land use, vehicle density, and primary roadways. Total number of vehicles considered here includes both transport and non-transport type i.e. two wheelers, four wheelers, taxies, auto rickshaws, buses, heavy vehicles (goods carriers and other transporters), tractors, trailers and others while the total number of industries presented includes both small-scale and large-scale factories in the city.

This is convinced that the major portions of the emission, transport and transformation of aerosols take place in the lower most part of the ABL, therefore the ACC retrieved for 200 m atmospheric layer adjacent to the ground is considered. Similar hypothesis could be found in Devara et al. (2002). The lidar-derived annual mean aerosol loading (column content) thus obtained and the number of vehicles available for the 14 years' period is overlaid on each other in Figure 15. It is clear that there is a considerable year-to-year variability in ACC that can be partly explained on the basis of varying aerosol generation/production, transport and removal processes around the experimental site. The increase in number of vehicles is slow initially almost up to 1991, and thereafter, it is very steep and peaked during 1994-1995, which is almost coincided with maximum ACC value. From 1995 onwards, the growth in vehicular density in Pune city showed high variability. More or less similar features can also be observed in the ACC. A high correlation (R^2 of 0.997) was obtained while comparing these two time series. But, a correlation coefficient of 0.40 was obtained between the number of vehicles and annual mean of ACC up to 1100 m AGL indicating that the local pollution events mainly affect the air quality of the lower most part of the boundary layer over the site. Figure 15 also illustrates variability of the yearly mean lidar-derived ACC and number of industries in and around the experimental station. Unlike the increase in number of vehicles, which is highly variable, the rate of growth of industries is almost monotonic, and shows some affinity with ACC.

This should be noted with sufficient care that the analysis performed here is preliminary one. However, the qualitative results show reasonable confidence on the influence of urbanization on the variability of the aerosol burden over Pune city. This could be further confirmed by analyzing high-resolution information about the urbanization parameters and other pollution related activities like transport of vehicles through the city, changes in the land-use pattern including building constructions, industrial developments etc.

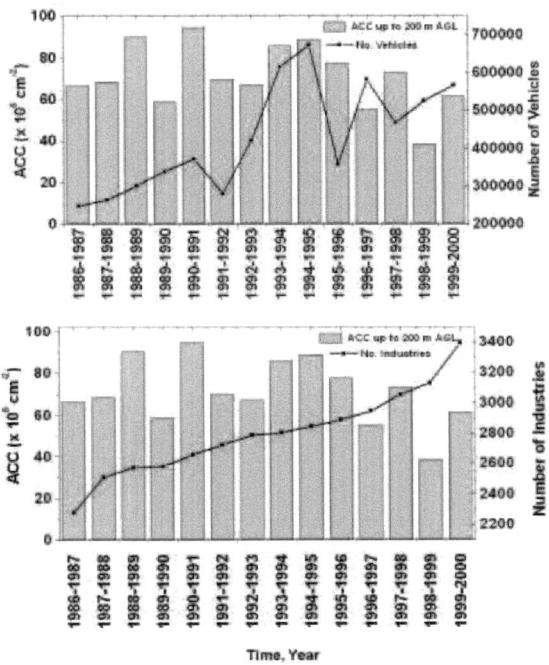

Figure 15. Co-variations between lidar derived ACC at 200 m AGL and number of vehicles (upper panel) and number of industries (lower panel) available in Pune city during the period between 1986 and 2000.

6. Conclusions and Future Scope

In this report, basically two efforts have been made, (i) to estimate and understand long-term changes in atmospheric aerosols over Pune, a fast growing urban station and (ii) to study the important and interesting properties of atmospheric aerosols and precursor gases on four clear sky days using compact solar Radiometer over Pune. The characteristics of aerosol those generated due to anthropogenic activities in the proximity of the experimental station and also those contributing to the local aerosol loading via long-range transport are examined. Another prime objective of the present work was to use the Wavelet Transform (WT) to the long-term ACC and AOD time series data to extract information that is not possible to unravel with a conventional Fourier or even Windowed Fourier transform. These studies gave a new insight into the long-term trends of aerosol loading in terms of aerosol dynamics.

This report detailed a unique data set of long-term observations of aerosols over a tropical station Pune, an urban site in western India. Not biased by limited amount of data set, this study investigated that the multi-year observations of lidar-derived ACC up to 1100 m and radiometer-observed columnar AOD showed an increasing trend. This increasing trend is supported by the growth of the industrialization, urbanization and land-use pattern changes in and around the measurement site during the study period.

Continuous wavelet-based spectral analysis on the time series of both lidar- and radiometer-measured aerosol data was performed to determine the periodicities present at different frequency ranges that are not possible to derive with traditional approaches like FFT. The resultant wavelet spectra yielded the occurrences of 12-14 months (AO) and around 22-25 months (QBO) periodicities at statistically significant level and their evolution in time with variable strength. These periodic features are further supported by the statistical significance test. The amplitude of QBO is found to increase with altitude and with decrease in wavelength. This event indicates that the observed QBO is of propagating type, and fine-mode particles are mainly responsible for the presence of such long-period waves in the variability of the aerosol loadings modulated by local meteorology over the experimental site (Devara et al., 1994).

In summary, multitudes of long term and short term modulations were found after the wavelet analysis on the time series. The results discussed could be an important source of information for the purpose of evaluation of the regional climate models which includes an explicit treatment of the aerosol-cloud interactions and aerosol indirect effect so that the role of aerosols on the different climate-relevant issues could be improved. This study could bring some more insights into the impact of increasing aerosol loading on rainfall. However, similar data set collected over different regions of the Indian subcontinent would bring more insight into this research activity. An assessment of the impact of urbanization on the aerosol loadings over the city was performed: high correlation coefficients between time variations in ACC, total number of vehicles and industries were found. The growth in population and associated urbanization also showed increasing trend as noticed in the long-term aerosol measurements presented in the study. These features are indicative of the influences and adverse effects of urbanization on aerosol loading over the city; consequently may have an impact on the local meteorology and environment. More detailed and quantitative insights into the role of urbanization on aerosol loading changes over Pune could be achieved by using high resolution (monthly) data sets consisting of the vehicle numbers, industries, population, detailed information on the traffic and transportation, domestic and industrial air conditioning and land use pattern changes at and around the city. Setting up of an integrated inventory dedicated to the archiving of above mentioned pollution related information for the city of Pune is considered to be very useful and an important task in this respect.

Bibliography

Andreae and Crutzen, 1996: *Atmospheric aerosols:* Biogeochemical sources and role in atmospheric chemistry, *Science,* **276**, 1052-1058.

Andreae, 1995: Climate effects of changing atmospheric aerosol levels –In: *World Survey of Climatology* **16**: Future Climates of the World. (Henderson- Sellers)- Elsevier, New York, 341-392.

Albrecht, 1989: Aerosols, Cloud microphysics & fractional cloudiness, *Science,* **245**, 1227-1230.

Angstrom, 1929: On the atmospheric transmission of sun radiation and on dust in the air. *Geograf. Ann. Deut.,* **11**, 156 - 166.

Angstrom, 1961: Techniques of determining the turbidity of the atmosphere. *Tellus,* **13**, 214 - 223.

Angstrom, 1964: The parameters of atmospheric turbidity. Tellus, **16**, 64 - 75.

Bolton et al.,1995:A Wavelet Analysis of Plio- Pleistocene Climate indicators, A new view of periodicity evolution, *Geophysical Research Lettter,* **22 (20)**, 2753-2756.

Carswell, 1983: Lidar Measurement of the Atmosphere. *Canadian Journal of Physics,* **61**, 378-395.

Chanin & Hauchecorne, 1991: Lidar Study of the Structure and Dynamics of the Middle Atmosphere. *Indian Journal of Radio and Space Physics,* **20**, 1-11.

Christoph and Grindewald, 2003: *Workshop on time series analysis of North Atlantic Oscillation.*

Chui, 1992: An introduction to wavelets. *Academic Press, Inc.,* Harcourt Brace Jovanovich, **266.**

Chui, 1992: Wavelets: A Tutorial Theory and Applications". Ac. Press, Inc. Boston.

Charlson et al., 1992: Climate forcing by anthropogenic aerosols, *Science,* **255** , 423-430.

Charlson et al., 1999: Direct Climate Forcing by Anthropogenic Aerosols : Quantifying the link between atmospheric sulfate and radiation- *Contr Atmos Phys,* **72**, 79-94.

Daubechies 1990: The Wavelet Transform Time-Frequency Localization of and Signal Analysis. *IEEE Trans.* Information. Theory. 36, 961-1004.

Dentener et al., 1997: Radiative forcing due to tropospheric ozone and sulfate aerosols, *Journal of Geophysical Research.* **102**, 28079-28100.

Devara et.al.1995: Aerosol profile measurement in the lower troposphere with four-wavelength bistatic Argon ion Lidar. *Applied Optics* 34,4416-4425.

Devara and Raj, 1989: Remote sounding of aerosols in the lower atmosphere Bistatic CW Helium Neon Lidar. *Journal of Aerosol Science.* 20. 37-44.

Devara et al., 1996: Investigations of Aerosol optical depth variations using Spectroradiometer at An Urban station, Pune , *Journal of Atmospheric Sciences,* **27** 621-632.

Devara et al., 2001: Some features of aerosol optical depth ,ozone and precipitable water content observed over land during the INDOEX-IFP 99. *Met. Zeirschrift,* **18**, 901-908.

Devara et al., 2001a: Lidar-radiometer study of boundary layer aerosols and their contribution to total column aerosol optical depth at a tropical station. In : 20th International Laser Radar Conference (ILRC) Book (Eds. A. Dabas, C. Loth and J. Pelon), 2001, 191-194.

Devara and Raj. 1987: A bistatic Lidar for aerosol studies. *Institute of Electronics & Telecommunication Engineers, Technical Review,* **4.** 412-415.

Devara, 1995a: Real Time monitoring of atmospheric aerosol using a computer controlled Lidar. *Atmospheric Environment,* **29**. 2205- 4425.

Erlebacher & Yuen, 2002: A Wavelet Toolkit for visualization and analysis of large data sets in Earth Quake Research. **Accepted** in the *Journal of Geophysics.*

Godtlibsen et al., 2003: Recent developments in statistical time series analysis : Examples of use in Climate Research.

Grossman and Morlet.,1984: Decomposition of Hardy function into square integrable wavelets of constant shape, SIAM. *Journal of Mathematics*, 15 (4), 723-736.

Hashvardhan, 1993: Aerosol Climate interaction. In Aerosol –Cloud-Climate interactions, Hobbs PV (ed.). *International Geophysics Series* .**54.** 75-95.

Hudgins et al., 1993: Wavelet Transform and Atmospheric Turbulence . *Phys. Rev. Letter,* **71**, 3279-3282.

Hofmann et al., 1993 Twenty years of balloon-borne tropospheric aerosol measurements at Laramie, Wyoming, *Journal of Geophysical. Research.* , **98** 12753-12766.

Hofmann and Solomon, 1989: Ozone destruction through.heterogeneous chemistry following the eruption of El Chichon, *Journal of Geophysical. Research*, **94** , 5029-5041.

Killinger & Mooradian, 1983: Optical and laser Remote Sensing, Springer-Verlag, Berlin.

Killinger & Menyuk, 1987: Laser Remote Sensing of the Atmosphere, *Science*, **235,** 37-45.

King et al., 1978: Aerosol size distributions obtained by inversions of spectral optical depth measurements, *Journal of Atmospheric Science*, **35**, 2153-2167.

Lau et. al., 1999: Climate Signal Detection Using Wavelet Transform- How to Make a Time Series Sing. *Bulletin of the American Meteorological Society,* **76**, 2391-2402.

Lelieveld et al., 1997: Chemical Perturbation of the Lowermost Stratosphere through Exchange with the Troposphere, *Geophysical Research Letter.* 24 , 603-606.

McClatchey et al., 1972: Optical Properties of the Atmosphere, *AFCRL–***72-0497**; Bedford.

Martin et.al.,1990: Glacial-interglacial CO_2 change: the iron hypothesis, *Paleo-oceanography* , **5 ,** 1-13.

Meyers et al., 1993: An Introduction to Wavelet Analysis in Oceanography and Meteorology with Application to the Dispersion of Yanni waves. *Monthly Weather Review.* **121,** 2858-2866.

Morys et al.,2001: Design, Calibration and Performance of MICROTOPS II handheld Ozone monitor and Sun Photometer. *Journal of Geophysical Research*, **106, D13** 14573-14582.

Morlet et al., 1982: Wave Propagation and Sampling theory- Complex signal and scattering in multi layered media, *Geophysics* **47**, 203-221.

Pandithurai et.al.1996: Aerosol size distribution and refractive index from bistatic Lidar angular scattering measurements in the surface layer. *Remote Sensing of the Environment.* **56** 87-96.

Perrier et al., 1995: Wavelet Spectra compared to Fourier Spectra. *Journal of Math. Physics* .**36**, 1506-1519.

Ramanathan et. al, 2000: The Indian ocean experiment: dispread haze from south and south-east Asia and its climate forcing, *Science* , submitted.

Sharma ,1994: Remote Sensing of the Atmosphere Using the Lidar Technique. *Ph.D. Thesis*, University of Pune, **167.**

Torrence and Compo. 1998: A Practical Guide to Wavelet Analysis . *Bulletin of the American Meteorological Society.* **79** No1. 61-77.

Twomey, 1974, Pollution and the planetary albedo, *Atmospheric Environment*, **8** , 1251-1256.

Yamamoto & Tanaka., 1969: Determination of aerosol size distribution from spectral attenuation measurements. *Applied Optics*, **8**, 447-453.

Zuev 1982: Laser Beams in the Atmosphere. Plenum press, New York.

Reports and Books

IGAP (The International Global Aerosol Program) Plan.1991. In Report of the Experts Meeting on Space Observations of Tropospheric Aerosols and Complementary Measurements, 51.

IPCC(Intergovernmental Panel for Climate Change) 1995. The Science of Climate Change, Houghton et al., Cambridge University Press., 572.

IPCC, *Climate Change, 1995 : The Science of Climate Change* , Cambridge University Press, 1996, ed., J. T. Houghton, L. G. Meira Filho, B. A. Callander, N. Harris, A. Kattenberg, and K. Maskell.

Statistcal Report, RTO, Pune (1980- 2002).

UNEP (United Nations Environmental Program, 2002). *The Asian Brown Cloud* : Climate and Other Environmental impacts, **53.**

Wavelets in Geophysics. Efi Foufoula- Georgiou . Praveen Kumar.

WCP (World Climate Program) 1983: In Report of the Experts Meeting on Aerosols and Their Climatic Effects, Deepak et al., WMO: Geneva., **107.**

WMO, 1991; Report of the Meeting of Experts to Assess Available Data and Define the Aerosol Component for GAW-BAPMON Stations (Boulder, 16 - 19 December 1991).

WMO, Global Atmosphere Watch Report Number 79. WMO Secretariat, Geneva, Switzerland.

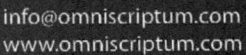

Printed by Books on Demand GmbH, Norderstedt / Germany